Crossfuze
Enterprise Services That Win

Table of Contents

Introduction

When you make the consequential decision to migrate enterprise services to ServiceNow, you understandably have tremendous responsibility resting on your shoulders. First and foremost, you want to get ServiceNow up and running, and you want to do it with speed, agility, and confidence. Along the way, you also want to put in place a ServiceNow implementation team, support this team with training and resources, and get a solid return on your ServiceNow investment. These technically oriented responsibilities aren't the totality of what's on your plate. You also need to transform your company's culture to support your long-term vision for a ServiceNow-powered digital transformation.

The reason behind ServiceNow's popularity is that it enables organizations to transform the organization from the inside out, touching on areas far beyond IT. ServiceNow does this by automating and streamlining the delivery and management of enterprise services, from HR onboarding functions to customer service to cybersecurity. Thus, as you work toward your long-term vision of a ServiceNow-powered digital transformation, you cannot focus your attention solely on managing the daily technical aspects of ServiceNow implementation.

The success or failure of your ServiceNow digital transformation initiative comes down to the people you will be counting on during your ServiceNow journey. Remember, digital transformation requires the support and buy-in of your entire business ecosystem—your employees, vendors, partners, and even your customers. Your ServiceNow user community will be called upon to learn new technology, adapt to new processes and workflows, and adjust policies and performance metrics. And even with these procedural changes, your organization will still only be scraping the surface of how to apply ServiceNow to accelerate digital transformation. Even more than the procedural implementation of change, your ServiceNow user community must buy into and believe in the benefits of change. They must understand change, demonstrate their commitment to support change, and, through their actions, move in unison toward change.

The complex work of leading widespread institutional change is known as organizational change management (OCM). It's as much an art as a science, and it involves equal parts hand-holding and hands-off facilitation. Successful OCM can separate the winners from the losers in any given market, yet achieving OCM success remains elusive for many organizations. A 2015 McKinsey study found that a whopping 70% of change management efforts fail, largely due to employee resistance and lack of management support. OCM is undeniably critical to driving the bottom line. A recent study by the Willis Towers Watson consulting firm found that companies that are effective at change management and communication are 3.5 times more likely to significantly outperform their industry peers that struggle in these areas.

Academic researchers and business leaders alike have extensively studied OCM, seeking to understand why it's important for achieving digital transformation and how to maximize its effectiveness. In this book, the *7 Pillars of Organizational Change Management to Achieve ServiceNow Wins*, Crossfuze has distilled decades of research, wisdom, and firsthand implementation experience into an easy-to-read, easy-to-navigate OCM guidebook.

This book deliberately does not dwell on theory and abstract academic concepts. Instead, we focus on the "how to" of change—that is, how to break down OCM initiatives into discrete, concrete stages and steps that are relevant to your management and decision-making responsibilities.

Companies that are effective at change management and communication are

3.5x

MORE LIKELY

to significantly outperform their industry peers that struggle in these areas.

A RECENT STUDY *from the* WILLIS TOWERS WATSON CONSULTING FIRM

Crossfuze has helped oversee more than 1,100 ServiceNow implementations, and we know what works and what doesn't in achieving change. This book will ensure you infuse your daily ServiceNow implementation-related responsibilities with a digital transformation mindset. As you read this book, we encourage you to begin developing your own action plan for how you will commit to implementing the guidance we present. This book's full value will only be realized when you put its ideas and insights into action.

If you find that you could benefit from some outside OCM consulting expertise, Crossfuze is just an email away at LetsTalk@crossfuze.com.

PILLAR 1

Establishing a Baseline Before Change

MEET ABIGAIL

Abigail is the vice president of business operations for a global manufacturing company. The five-decade-old company has been through several technology-driven efficiency initiatives already that have transformed key areas, including timekeeping, payroll, and the sales funnel. These change initiatives did not always go smoothly, but the company eventually adapted, and Abigail doesn't hear anyone talking today about the rough patches. Now the C-suite is getting ready to greenlight a comprehensive modernization of the customer experience using ServiceNow's Customer Service Management (CSM) platform. CSM will touch almost every service unit in the organization, from sales and marketing to accounting and legal. Abigail is tasked with leading this Customer Experience (CX) modernization, but she is concerned that the company isn't ready to embrace a change of this magnitude. She knows that many CX leaders think the company is already taking advantage of next-generation CX tools via omnichannel strategies. But these leaders don't think about how to connect and integrate the organization's CX tools, nor is Abigail confident they would be enthusiastic about the ServiceNow digital transformation initiative. Abigail decides that the first step to organizational change management (OCM) is to conduct a baseline assessment to understand the organization's readiness for change.

Abigail's concerns about change readiness are commonplace in industries of all types and sizes. A 2017 Deloitte study found that only 8% of businesses report having access to usable data on change readiness in their organization, and only 9% report having a strong understanding of how human talent drives performance in their organizations. The reality is, too many organizations are rushing into change, blithely unaware that change readiness is key to eventual OCM success. A change readiness assessment is an opportunity to establish a strategic baseline of an organization's ability to achieve change. By getting a clear picture of what your company looks like today, you become better-positioned to appreciate the shortcomings of the

A change readiness assessment is an opportunity to establish a strategic baseline of an organization's ability to achieve change.

status quo, and to gain clarity around your vision for change. Furthermore, your baseline will inform the development of a roadmap that sets up a logical timing and pacing for change.

When implementing a technology solution as elegant and game-changing as ServiceNow, it's critical that organizations do sufficient legwork to understand employees' attitudes and aptitude for change. These insights ensure the organization can remain relentlessly focused on taking advantage of all aspects of ServiceNow to deliver measurable financial value, minimize risk, and enable and empower users to work better.

Let's explore the key elements necessary to conduct a change readiness assessment that will set you up for optimal OCM outcomes.

1.1. TARGET EMPLOYEES WITH STRONG VIEWPOINTS

To ensure you gain a comprehensive perspective on who supports change, you want to ask your management team for input on which employees are most uncomfortable with change. The concerns voiced by these employees will become the foundation of your strategy and pacing for implementing ServiceNow. These employees not only tend to have strong opinions on change, but they tend to want to influence others' attitudes toward change. By spending time with them and asking for their feedback, you are showing them that they have some level of control over the changes that will accompany ServiceNow digital transformation. This perceived control will help these employees feel less stressed and make them more receptive to change—rather than fight you every step of the way. In particular, one-on-one interviews are essential for drawing out depth and nuance in people's feelings and opinions about change. Key themes to cover in interviews include the following:

- How well do you understand the rationale for change?
- How well do you understand the benefits of change?
- How well does the organization do in disseminating this information to stakeholders?
- How much appetite is there for change in the organization?

1.2. PINPOINT WHERE YOUR GREATEST POTENTIAL PITFALLS LIE

As you develop a sense of your organization's readiness for change, you want to use these insights to predict where in the process of change your organization is most likely to stumble and fall. For example, if you get the sense that middle management in your organization often fails to lead by example, you can target your middle managers with improved training and mentoring, or perhaps revise job performance expectations to more clearly emphasize the importance of leading by example. The more potential pitfalls that you can identify and nip in the bud before you initiate ServiceNow implementation, the smoother the change will be.

1.3. RECOGNIZE THAT LEADERSHIP MUST HAVE CHANGE CAPACITY AS WELL

An organization's front-line and mid-level workers aren't the only ones who should be the focus of change readiness assessments. The equation is more complex, as you also need executive leadership to possess a high capacity for change through talk and action. A book published by the Association for Supervision and Curriculum Development recommends that organizations think of change readiness as a matrix, where employees' capacity for change is plotted on the x-axis, and leadership's capacity for change is plotted on the y-axis.

What this matrix illustrates is that readiness for change is achieved only when both employees and leadership demonstrate a high capacity for change. If only leadership has a high capacity for change, the organization is ready to learn but not to change. Conversely, if only employees have a high capacity for change, the organization is opening itself up to frustration. And if neither leadership nor employees have a high capacity for change, any change initiated within the organization will encounter open resistance, including from middle managers who get into turf wars because no one is aligned with or prepared for change.

1.4. DON'T OVERTHINK OR OVEREXTEND THE REACH OF A CHANGE ASSESSMENT

A change assessment is just that—a baseline evaluation of your organization's attitudes toward change. It's not intended to serve as a be-all, end-all template for change, nor should it evolve into a problem-solving exercise. The reality is that you will never have time to amass every possible bit of information to assess your organization's ServiceNow readiness, nor should you waste time trying. Likewise, in the process of soliciting feedback about the state of change readiness in your organization, you don't want to become derailed trying to immediately solve the problems you identify. The more you get bogged down before implementing ServiceNow, the less efficiently you'll be able to start winning with ServiceNow.

1.5. BUILD CONSENSUS AROUND YOUR BASELINE ASSESSMENT

As you tabulate and analyze your assessment results, key themes will begin to emerge about where you are and where you need to be. Your focus should be on building consensus around these findings. Once everyone agrees on the baseline, you can all work together to address the problems you've identified and optimally position the organization for success with ServiceNow. Many companies set up a formal change assessment committee that reviews outcomes and then makes binding decisions about how to proceed to address any shortcomings. A committee that represents all of the stakeholders responsible for implementing ServiceNow will be much more likely to be responsive to group decisions than top-down mandates. For example, if more ServiceNow training or outreach initiatives are needed, the committee is much more likely to make those things happen than a single management executive could.

1.6. ENSURE PLANS DEVELOPED FROM YOUR BASELINE ASSESSMENT ARE FLEXIBLE

As you develop plans to remedy the issues identified in your baseline assessment, you want to keep these plans agile and open to adjustments based on new information. No one should expect that these plans will represent the totality and finality of your ServiceNow implementation journey. Instead, you want to create workflows and processes that encourage revisiting plans and making adjustments in an orderly, transparent fashion.

Key Talking Points to Cover with Your Colleagues

Modern organizations are increasingly feeling pressure to accelerate the pace of change. Product lifecycles are increasingly compressed, and customers are demanding more streamlined and automated services. To manage change effectively, the key first step is to understand where your organization is in terms of readiness for change.

- An effective change readiness assessment should target employees with strong viewpoints about change.

- Leadership must have change capacity as well and must be careful not to overthink or overextend the intended purpose of change assessments.

- As you develop a game plan for optimizing change readiness, you want to ensure it remains mutable and flexible.

- Change readiness assessments can serve as a springboard for educating employees about the goals of change. You can share with employees what transformative change will look and feel like, and how it will be beneficial to the employee experience.

PILLAR 2

Developing an Action Plan for Change

MEET NICHOLAS

Nicholas is the CIO of a fast-growing networking technology company who has been given a mandate to reduce IT costs by 10%. Because the company has grown so unpredictably fast, IT services are delivered in a piecemeal fashion, with multiple help desk applications that don't talk to one another and a custom-built IT management platform that is poorly integrated. Nicholas recognizes that ServiceNow is the obvious choice for unifying and connecting all IT service management functions across the organization. But he also recognizes that ServiceNow is not a "rip-and-replace," "lift-and-shift" proposition. In other words, mid-level IT managers in his organization won't want to do the hard work of reimagining and redesigning workflows to fit ServiceNow best practices for IT Service Management. Instead, these IT managers will instinctively take the path of least resistance and try to force ServiceNow to conform to their outdated, inefficient workflows. Fully aware that this will negate much of the benefit of migrating to ServiceNow, Nicholas knows he needs to put in place a rock-solid plan of action—one that will position everyone in his organization to make the transition to ServiceNow properly. Nicholas's action plan cannot just include a roadmap for implementing the technical aspects of ServiceNow; Nicholas also recognizes that he must develop a plan for driving changes in how managers think about the delivery of IT services.

Nicholas's conscious efforts to develop a plan for change management in his IT department is the right move. A 2018 survey of IT project leaders by the Axelos consulting firm found that project success is four times higher in organizations where project leaders understand business strategy and have aligned themselves with this strategy. Organizations instinctively blame technology when a major investment such as ServiceNow has failed to meet expectations. In reality, it's the people, processes, and planning that spell the difference between ServiceNow success and failure. That's why it's so important that every organization has a plan of action for supporting and

driving organizational change. As you look to develop your ServiceNow implementation success plan, let's explore the most important elements to keep in mind.

2.1. PRIORITIZE THE CAPABILITIES YOU MOST WANT TO BUILD

People are most likely to support changes that directly solve their pain points. To ensure your action plan is responsive to the concerns of your target audience, you want to work with this audience to prioritize the ServiceNow capabilities and features you intend to build. Simple heat-mapping techniques (high, medium, low) will help everyone identify and agree on the most critical gaps and shortcomings. You then want to build your implementation around these identified priorities. Your target audience will appreciate that you are solving their biggest problems, which, in turn, will promote stronger buy-in and support for your implementation plans over the long term.

2.2. ESTABLISH SUSTAINABLE PACING

One of the most important decisions you'll make as you design your ServiceNow implementation roadmap is to set a sustainable pace. Although you may be inclined to go as quickly as possible, your audience may not be able to keep up; if they become confused and disengage, your implementation plan will fall apart. Conversely, if you don't move fast enough, you could disappoint your target audience and cause them to lose faith and/or interest in your plans. You want to develop your ServiceNow implementation roadmap in consultation with your stakeholders. When they buy into your roadmap, they'll be much more likely to help keep you on schedule.

2.3. DEVELOP A GOVERNANCE FRAMEWORK TO GUIDE DECISION-MAKING

As you dive into the day-to-day of ServiceNow implementation, you'll quickly realize how many small but consequential decisions need to be made. The worst outcome of these decisions would be if you were to alienate and/or upset your end-user stakeholder community. If they feel that the decision-making process wasn't transparent or consistent or fair, they will be less likely to buy into and support the overall ServiceNow implementation plan, and more likely to actively petition to get decisions reversed and modified. A governance framework that clearly lays out the process by which you'll make decisions is crucial to combating this adverse outcome. Your governance framework should include a decision tree that describes your standard decision-making process, a hierarchy that explains who will have decision-making authority, and predetermined courses of action that the organization will commit to taking based on various potential scenarios. The more fleshed out your governance framework is, the less likely anyone is to be caught off guard by the decisions that get made, or to campaign to reverse these decisions.

2.4. INTEGRATE GUARANTEED EARLY WINS TO BUILD MOMENTUM

The beginning of your ServiceNow implementation journey shouldn't be hampered by difficult projects and a long road to achieving your first wins. To maintain people's investment and focus on the task at hand, you want to build in guaranteed early wins. These are the relatively low-hanging fruit that you can easily pick off; they will give everyone a confidence boost and inject forward momentum into all of the tasks to come. For example, when it comes to IT Service Management implementation, ServiceNow recommends that you focus initially on high-visibility, low-cost opportunities that quickly engage a large base of users in requirements development, acceptance testing, and usage. These quick-win opportunities include request fulfillment, incident management, and a service catalog.

You must recognize that unforeseen obstacles and other challenges are inevitable.

2.5. CREATE PROTOCOLS FOR OVERCOMING OBSTACLES

As you develop your implementation plans for ServiceNow, your goal should not be to create a seamless, obstacle-free implementation. To the contrary, you must recognize that unforeseen obstacles and other challenges are inevitable. Your best defense is to develop protocols for how you'll efficiently work through these challenges. You should start this process by making a list of some typical obstacles that might arise. Then, these tangible examples can serve as the foundation around which you build protocols for overcoming these obstacles.

2.6. DEVELOP A PLAN FOR MONITORING AND REPORTING PROGRESS

Every project undertaken by an organization should include built-in ways to monitor and report on progress. This adage is particularly true when considering that people respond positively to progress (including your silent cynics) and are motivated by progress. You want to capitalize on this during ServiceNow implementation by measuring progress on a variety of fronts—and share this progress with others. Key questions around which to build these progress assessments include the following:

- How closely are we adhering to our ServiceNow implementation plans?
- How closely are we adhering to our roadmap timeline?
- How ready are employees to support the change we're seeking?
- How is productivity improving?

Key Talking Points to Cover with Your Colleagues

Your action plan to support a successful ServiceNow implementation is about so much more than just the technical aspects of implementation. To properly position ServiceNow for long-term success in your organization, you want to work proactively and strategically to sustain the organizational change you seek.

- It's your job to prioritize the capabilities your organization most wants to see developed; that's how you'll get the biggest bang for your buck.

- You want to establish a pacing that is neither too fast nor too slow, as well as set up a governance framework to support transparent, consistent decision-making.

- You will need guaranteed early wins on your ServiceNow implementation journey to supercharge your forward momentum.

- Don't forget to develop plans for overcoming anticipated obstacles and for systematically monitoring your progress.

Learn how to answer employee questions about digital transformation like the ones below by reading the article "3 questions from employees on digital transformation and how to answer" from CIO.com.

1. Will I have a job after tasks are automated?
2. How can I contribute to the company's goal to transform?
3. How can I keep pace with all the new technologies?

PILLAR 3

Harnessing Employee Insights Effectively

MEET JACK

Jack is the executive vice president of employee relations for a Fortune 500 telecommunications company preparing to migrate 118,000 associates worldwide to ServiceNow's Customer Service Management (CSM) platform. CSM will transform the customer experience from the inside out, connecting previously disconnected systems and requiring sales, marketing, and customer service teams to collaborate together like never before. Jack and his C-suite colleagues have been working for months to plan for implementation, and Jack has been leading the development of employee training programs to ensure a smooth transition. But when Jack launched an initial round of presentations to explain the pending migration to ServiceNow CSM, employees began feeling anxious and worried. Many of them complained that the new workflows under CSM will be too complicated and inefficient for day-to-day use. Some even suggested that ServiceNow CSM should be implemented based on the company's current customer service-related workflows—instead of leveraging the CSM best practices built into the platform. Furthermore, at all of the introductory presentations, employees have already begun offering their own ideas, suggestions, and tips for how to tweak and improve the new CSM system and its many interconnected workflows. Jack knows he wants to collect all of this employee feedback in a systematic, effective way, but he doesn't know how to do it—or how he can make time to do it.

At the heart of Jack's concern is a universal constant in organizations preparing for large-scale change. The act of listening to employees—genuinely listening and taking feedback to heart—is what engages employees and motivates them to support organizational change. A 2017 McKinsey & Company study found that companies experiencing success with digital transformation are nearly 60% more likely to reporting having visibly engaged front-line workers than companies that report struggling with digital transformation. For employees to buy into and embrace seismic shifts in how they do their jobs, they need to believe that their employer cares about their fears and criticisms. Similarly, companies will always be more successful when they systematically gather employee feedback.

Employees provide key feedback of all kinds. They can help identify flaws in workflows, pinpoint challenges with day-to-day implementation of challenges, and suggest strategies for weeding out waste and inefficiency. Let's explore the most essential strategies for harnessing the insights, perspectives, and attitudes of your employees to successfully drive the organizational change you seek.

3.1. DEVELOP SURVEYS TO GAUGE EMPLOYEE ATTITUDES ABOUT CHANGE

When you solicit feedback from employees who are concerned and apprehensive about ServiceNow implementation, you're likely to start getting more feedback than you know what to do with. You can quickly become overwhelmed trying to synthesize and make sense of this feedback.

How do you systematically compile and analyze all of your employees' insights? How do you ensure you're truly understanding how employees feel, and not skewing your understanding based on personal bias and/or non-representative sample sizes? One of the most effective ways to move past qualitative, subjective impressions is to develop quantitative surveys that measure employees' attitudes and aptitude for change. For example, you could develop a survey asking employees to rate their level of agreement with statements on a standardized 1-10 sliding scale. These statements could include the following:

» Past technology upgrades in this organization have been successfully implemented.
» This organization has effectively communicated the rationale behind moving to ServiceNow.
» This organization has effectively prepared employees to use ServiceNow.
» I can articulate the value of migrating to ServiceNow.
» Middle management is a champion of ServiceNow.
» Executive management has developed an effective strategy for implementing ServiceNow.

When you average employee responses to survey questions, you will be able to quickly get a comprehensive snapshot of what issues concern employees the most about ServiceNow, which types of employees and departments need more support and resources, and how effectively you are rolling out ServiceNow and using it to spark meaningful change. You should consider rolling out these surveys at multiple points during ServiceNow implementation—before, during, and after major milestones of your long-term implementation journey. Not coincidentally, inviting your employees to take these surveys also helps them understand and mentally prepare for change; they will recognize that change is coming and that they can have a say in it.

3.2. FOCUS ON EMPLOYEES WHO WILL BE MOST IMPACTED BY CHANGE

An organizational change initiative like ServiceNow will affect various types of employees and service units differently. Your goal should be to communicate and engage with them all. But at the same time, you also want to prioritize your efforts based on who will be impacted most. This process starts with identifying and ranking the types of employees and departments within your organization who are most likely to be affected by migrating to ServiceNow; this is your priority list that you want to keep in mind as you develop all of your communications and outreach initiatives. In fact, if you don't include these people and make them feel that they have some control over the change, they become likely to undermine your plans for ServiceNow implementation. Keep in mind that many of the employees who will be most impacted by change feel that they have little say or control over what happens; this is particularly true of your front-line employees. Thus, you want to look for small ways to help them feel a bit more in control during ServiceNow implementation. For example, perhaps you could alleviate some micro-managing that they typically endure, or perhaps you could shift some approvals processes down the chain of command to reduce the intimidation they feel getting these approvals.

3.3. PROACTIVELY WORK TO CAPTURE THE VOICE OF THE SILENT MAJORITY

During any change initiative, the loudest, most vocal critics and skeptics tend to drown out the voice of all others. This can skew your impressions of how employees feel about and respond to ServiceNow; it also can result in the development of implementation plans that aren't optimized for or embraced by the silent majority. The solution to overcoming these challenges is to proactively work to capture the voice of the silent majority. Instead of exclusively relying on town hall-style listening sessions where the most vocal tend to dominate the conversation, you also want to pursue strategies such as setting up an online portal to serve as a suggestion box or inviting handpicked employees to focus groups and one-on-one interviews. Finally, you want to ensure you're not allowing the perspectives of any particularly vocal individual to dominate the conversation around ServiceNow implementation. One particularly effective strategy is to pull aside your most vocal critics and have a heart-to-heart with them. Sometimes these critics will end up backing down or formally retracting some of the harshest things they've said, especially if you can help them realize their thoughts are quite far out of line with how the silent majority feels.

Pull aside your most vocal critics and have a heart-to-heart with them.

3.4. CREATE SOCIAL EXPERIENCES THAT ENCOURAGE OPEN, FREE-FLOWING DIALOGUE

One of the key reasons for the meteoric rise of online forums and social media is that they're inherently and universally engaging to people. These online social experiences also serve another purpose: The postings become invaluable for gauging people's reactions to things and ideas. You want to foster this open, free-flowing dialogue around your ServiceNow implementation journey so that you can maintain a pulse on how employees are reacting and responding to change. For example, you could host internal "jam sessions" and "blitzes" that help raise the visibility of various stages of ServiceNow implementation. These engaging social experiences also can go a long way toward reinforcing for employees that their participation is valued and their feedback matters.

3.5. LOOK BEYOND WHAT EMPLOYEES SAY; STUDY WHAT THEY DO

While there's no question you must listen carefully to what employees tell you about ServiceNow, you don't need to limit yourself to taking your employees' word at face value. In the world of analytics, tracked behaviors often provide more accurate and meaningful insights than do verbal assessments in terms of understanding how employees are responding to change. It's not that employees would deliberately lie to you, but sometimes observed insights will outright contradict statements from employees or put what they've told you into an entirely different light. Even if direct observations merely confirm what your employees already have told you, you'll gain increased confidence that you've captured employee sentiments accurately, and be able to come from a position of strength as you advocate making course corrections and other changes to how you implement ServiceNow.

3.6. CIRCLE BACK TO SHOW EMPLOYEES YOU LISTENED

No matter how you gather feedback and insights about ServiceNow from your employees, you want to circle back by sharing with them what you learn in aggregate. In fact, you ideally want to show your employees how you made specific changes in response to the insights you gleaned from them. For example, if you conduct a survey one week prior to your Phase 1 go-live date and you realize your employees desperately need a ServiceNow refresher tutorial delivered via webinar, you don't just want to schedule the webinar; you also want to show employees the results of the survey, and explain to them how you interpreted those results and ultimately decided to schedule the additional webinar. Not only does this level of transparency engender goodwill among employees, but it also helps employees to more fully appreciate why an additional level of support is being provided to them.

Key Talking Points to Cover with Your Colleagues

Your employees are on the front lines of any change initiative in your organization. Their success adjusting to change becomes your success. Conversely, if they struggle to adapt to ServiceNow, you will struggle to achieve the enterprise-wide ServiceNow wins you seek.

- Your employees are your most powerful assets at every stage of your ServiceNow implementation journey.

- You want to be working alongside your employees to harness their insights, ideas, and perspectives, particularly the employees who will be most impacted by change.

- You can use a variety of approaches to maximize your effectiveness in gathering employee feedback.

- Surveys that gauge employee attitudes and aptitude for change are essential in providing accurate, quantitative perspective regarding the changes you seek.

PILLAR 4

Effectively Communicating Change

MEET NORA

Nora is the CIO of an aerospace manufacturing company that is preparing to migrate Customer Service Management (CSM) and Project Portfolio Management (PPM) operations to ServiceNow. The decision to move these services to ServiceNow was a logical one for the company. Five years ago, the company dramatically streamlined and automated its IT Service Management functions using ServiceNow. However, Nora is anticipating a tougher hill to climb with CSM and PPM. She's already hearing rumblings at the department head level about how illogical and ill-timed the transition seems. Although Nora's company already has announced that the migration to ServiceNow CSM and PPM will not even begin for six months, employees are openly expressing criticism about the upcoming transition and spreading misinformation and rumors about how employees will be affected. Nora knows that if she doesn't get ahead of these challenges soon, the rumor mill will spin so out of control that she may never be able to get her ServiceNow initiative back on track. Nora talks over the situation with her company's ServiceNow implementation consultants and learns that her top priority should be to develop a comprehensive, encompassing communications strategy. This communications strategy must directly involve every leader and manager in the organization and must be operationalized as soon as possible.

The pressure that Nora feels to develop a communications strategy for managing ServiceNow-related organizational change is a universal experience. A McKinsey study from 2015 found that when companies mount an effective communications strategy for sharing progress on organizational transformation initiatives, employees are eight times more likely to report a successful transformation than when this communication doesn't occur. Migrating enterprise services to ServiceNow might seem like primarily a technical challenge, but the reality is managing the technical side is relatively simple and straightforward. An organization's biggest challenge by far is communicating the value proposition of change to skeptical employees, convincing employees to give up deeply entrenched ways to get work done, and transitioning employees to new workflows and new systems.

To sell employees on ServiceNow-powered digital transformation, it's your job to explain how a platform that delivers and manages enterprise services can serve as the spark and accelerator of enterprise-wide digital transformation. There is a tremendous amount of nuance and detail to unpack here. Let's explore the key building blocks for developing a comprehensive plan for effectively communicating ServiceNow change to employees.

4.1. FLESH OUT YOUR CORE MESSAGES

Before you can share information about ServiceNow implementation with your employees, you must understand why you're going to all the trouble of implementing ServiceNow in the first place. This process starts with defining the core messages you wish to convey. The core messages have little to do with the procedural and technical aspects of

8X

When companies mount an effective communications strategy for sharing progress on organizational transformation initiatives, employees are eight times more likely to report a successful transformation than when this communication doesn't occur.

2015 McKINSEY & COMPANY STUDY

ServiceNow implementation, and everything to do with the basic "whys" of change. They're questions that every employee will be curious about, and that are essential to convincing employees to support and buy into your ServiceNow implementation vision.

Before you start communicating the "whys" of change, it's important that you challenge yourself to articulate eloquent answers to these questions. On an intuitive level, you no doubt understand the "whys" of change, but it is actually quite difficult to convert your thoughts into cohesive, convincing statements that will resonate with employees. To prepare eloquent responses to the "why" questions, ask yourself the following questions:

- Why do we need to make the change right now?
- What happens if we don't make the change right now?
- Why is the existing solution no longer sufficient?
- Why will our organization be better off with this change?
- Why should this change be viewed as complementary to our previous change initiatives?

The answers to these questions will be your core messages, the ones that will be at the heart of—and the driving force behind development of—your communications strategy.

4.2. GET EMPLOYEES INVOLVED FROM DAY 1

Many organizations feel the need to "get their ducks in a row" before publicizing their ServiceNow implementation plans to their broader employee base. But employees don't see it this way. They don't like feeling that the wool has been pulled over their eyes. The reality is that it is never too early to begin communicating with employees about change and engaging them in planning for change. They will appreciate the gesture and will respond in kind, becoming more involved and more engaged with ServiceNow.

To sell employees on ServiceNow-powered transformation, it's your job to explain how a platform that delivers and manages enterprise services can serve as the spark and accelerator of enterprise-wide digital transformation.

4.3. COMMUNICATE CONTEXT AND DESIRED OUTCOMES, NOT TASKS

Senior leaders tend to break down abstract, big-picture goals and priorities into concrete, actionable task lists. The goal of these task-oriented projects is to provide everyone under them with a series of clear, discrete steps for achieving big-picture change. Senior leaders genuinely believe they're doing everyone a favor; after all, they are eliminating uncertainty and ambiguity with assignments while making work more enjoyable, right?

The problem is that the big-picture rationale and context for change get lost with this task-oriented approach. Although it may seem counterintuitive, you're actually better off communicating context and desired outcomes, not tasks. Your employees will surprise you. Not only will they instinctively come together to figure out what tasks they need to do to achieve a specified outcome, but employees will also develop solutions, ideas, and workflows that are better than those you would have imposed on them. There's never any harm in communicating context and desired outcomes; even if you think your employees will struggle with this approach, you can always outline a suggested roadmap of tasks as well. Over the long term, remember that this is a key opportunity to nurture the growth of the next generation of leaders in your company—the ones you're preparing to help you solve future challenges.

4.4. SHOW HOW THE CHANGE IS BENEFICIAL AT MULTIPLE LEVELS

The first question any employee will ask about a change initiative is, "What's in it for me?" You absolutely must be prepared to deliver a compelling answer customized for the type of employee who's asking. But beyond that, you want to show how the change initiative will benefit people at every level of the company—the team, the type of enterprise service, the company, and, ultimately, the customer. The most effective way to communicate these benefits is to tell compelling stories about change, tailored to specific target audiences. With each story you tell, you want to contrast a specific "before" with a specific "after," so the change becomes tangible and concrete to your target audience. As the saying goes, "Don't tell me—show me!"

BEFORE **BENEFITS OF CHANGE** AFTER

TO INDIVIDUAL EMPLOYEES

TO YOUR TEAM

TO THE WAY SERVICES ARE DELIVERED

TO THE COMPANY

TO THE CUSTOMER

4.5. TARGET SPECIFIC AUDIENCES WITH TAILORED MESSAGING

As you look to publicize your ServiceNow implementation roadmap, your aim shouldn't be to develop a single standardized presentation to all of your stakeholders. Your implementation plans will affect various stakeholders very differently, and disparate stakeholders will care about very distinct aspects of the implementation journey. The more you customize your messaging to multiple audiences, the more your messaging will resonate with them, and the more they'll buy into your ServiceNow vision. Tailored messaging also is important to combat the skeptics and naysayers within certain departments or business areas who have a very specific set of concerns or criticisms of your plans; you want talk in language that will resonate with them.

4.6. BRAND THE CHANGE

You could call your change initiative "ServiceNow" or "Upgrade to ServiceNow," or you could give it a catchy, strategic name that offers a distinct brand identity, such as "OneNow" or "CustomerEdge" or "Reinventing Customer Service with ServiceNow." By branding your ServiceNow initiative, you are ensuring that employees see the change as not just another routine technology upgrade, but as a strategically important, discrete initiative that helps the customer. You want employees to be proud of this initiative and to rally around it. Additionally, the brand name for this successful project will raise its profile and help it shine on the resumes of the team that drove the change.

4.7. BE TRUTHFUL ABOUT CHANGE

Many organizations fear that employees "can't handle the truth," so they devise all sorts of strategies for obfuscating, sugarcoating, and spinning the truth to make it more palatable. The worst thing that you can do when discussing a major change initiative is to create the appearance that you're trying to hide something. The more upfront and forthcoming you are, the less likely employees are to feel skeptical, resentful, and even deceived by the change. Similarly, the more opportunities your employees have to be involved from the beginning, the more transparent the change will feel to everyone. Being truthful about change also extends to the origins and impetus for change: if the change is being mandated from up high, you shouldn't imply it's an organic, grassroots initiative. Similarly, if the change doesn't yet have widespread buy-in and support, you shouldn't suggest that everyone is already on board.

> "Honesty always pays. Honesty alone will never build a business, but the policy of honesty, of scrupulous integrity, will—other things being reasonably equal—always win in the race for success."
>
> HARRY GORDON SELFRIDGE | FOUNDER, SELFRIDGE'S DEPARTMENT STORE

4.8. END WITH A CALL TO ACTION

Change is tough for employees to absorb and comprehend, especially at the beginning when it's just an abstract concept that lives in a PowerPoint presentation. You may think you've been effective in walking employees through your change initiative, but employees often absorb only a small fraction of the information. To ensure you end on a strong note, you want to provide a simple takeaway message in the form of a call to action. A call to action is a strategy for telling your audience exactly what to do next. It should be straightforward and direct, such as "We invite you to watch the three-minute Introduction to ServiceNow video that's been posted to the company intranet homepage."

4.9. OVERCOMMUNICATE CHANGE

In any organization, employees are being constantly bombarded with performance reports, status updates, email reminders, and internal admin requests—all of which come on top of an employee's primary job responsibilities. Your ServiceNow digital transformation initiative is competing for mindshare with very busy companies, which means it's imperative that you overcommunicate change. You want to repeat key messages often, and use a variety of multimedia—from graphics and PowerPoint decks to videos to social media and text messages—to convey the most important information about change. Redundancy and repetition are your best friends.

Key Talking Points to Cover with Your Colleagues

As important as effective communication is during change, there is not one single, perfect way to communicate change. Change can be uncomfortable and difficult, both for the senior leaders who must communicate change and for the employees who must adapt to change.

- The value proposition of ServiceNow comes down to explaining how the platform can spark and accelerate digital transformation across your organization.

- Before you begin communicating the "whys" of change, you want to challenge yourself to develop compelling, clear answers that will resonate with employees.

- You can help employees relate to change by communicating in outcomes instead of tasks and telling engaging stories about the benefits of change.

- Developing an internal plan for marketing ServiceNow to employees will go a long way toward helping them identify with ServiceNow.

PILLAR 5

Taking Advantage of Change Agents

MEET CAL

Cal is the executive vice president of business operations for an automotive parts supplier with factories and offices in seven countries. The automotive sector is a fast-paced, ultra-competitive industry; Cal and the rest of the leadership team recognize that unless the company continually innovates and evolves, it will quickly be overtaken by the competition. Cal's company already has begun implementing ServiceNow's Customer Service Management (CSM) app to improve the customer experience, particularly order management and customer communications. CSM implementation is having a ripple effect on workflows and service units across the company. Teams are being told the way they've been doing things is no longer good enough, that there will need to be a higher level of thought and coordination that goes into their daily work. At the macro scale, Cal recognizes that these changes will dramatically enhance the customer experience. But on the microscale, employees already have been raising concerns and dragging their heels with implementation. Cal and his C-suite colleagues have rolled out numerous training initiatives to provide some much-needed hand-holding. And the good news is that Cal is receiving reports that some individual employees are embracing the changes and becoming natural leaders in helping others through the transition. Cal knows that these individuals are serving as strategic change agents for the CSM initiative, and he wants to put in place systems and processes for identifying and mentoring more of these change agents. What should Cal be doing to empower his change agents to help drive his CSM digital transformation initiative?

Every organization has individuals just like the change agents at Cal's company who intuitively understand how to use technology to drive digital innovation. They are people who possess a unique combination of credibility and influence among their peers that enables them to be highly effective at driving organizational change. A recent Deloitte survey found that 88% of

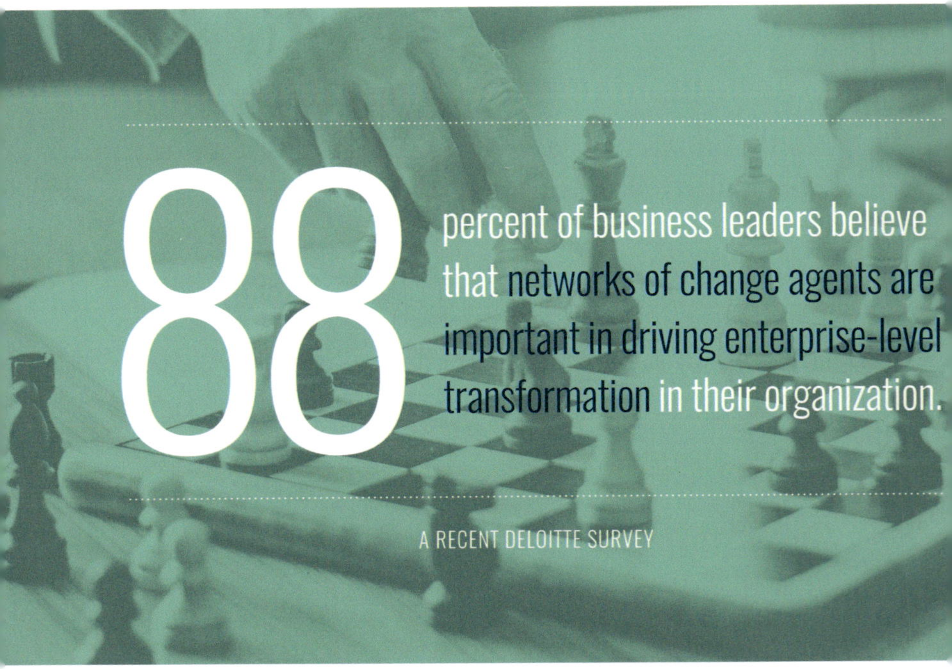

business leaders believe that networks of change agents are important in driving enterprise-level transformation in their organization.

In many ways, change agents are the unsung heroes of digital transformation: they work behind the scenes to position their organization for success and to help their peers adapt to change. Change agents stand apart from their peers because they

- work without an official mandate, outside their formal job responsibilities;
- rely on very organic, grassroots processes to drive change;
- excel at building camaraderie, rapport, and mutual trust with peers;
- are completely comfortable around new technology;
- are early adopters of digital trends;
- are passionate and determined; and
- truly believe they can contribute meaningfully to change.

You can ask supervisors and managers at all levels of your organization for help identifying these employees; your HR department also will have ideas and suggestions for identifying your change agents. Let's explore the most important things you can do to take full advantage of change agents to drive the digital transformation you seek.

5.1. CREATE A CULTURE THAT NURTURES AND ENCOURAGES CHANGE AGENTS

Most employees do the jobs that are assigned to them and, even when they excel, won't step outside of their box. Change agents are the rare employees at all levels of an organization who intuitively identify pain points and then develop effective solutions for them, regardless of whether it's logically part of their formal job responsibilities. Change agents are constantly conversing with peers, absorbing information, making mental connections among disparate ideas, and thinking up new ways of doing things. To effectively encourage these sorts of behaviors within your organization, you want to do the following:

» Instruct the supervisors of change agents to meet with these folks regularly to get feedback on how change is progressing, and to flesh out and articulate their ideas for improvement.
» Formally grant change agents time during their workday to help drive change; they typically would be squeezing this work into their schedule on their own, so imagine how much more effective they will be if time is set aside for them to work their transformative magic.
» Introduce change agents to other change agents in your organization to form strong change networks.
» Encourage employees to form advisory committees and workgroups to develop even more ambitious and coordinated programs to help drive change.

- Create a "job description" that defines what a change agent is within your organization and the expectations for the role; then, you can privately pull aside change agents and encourage them to "apply" to these informal roles.
- Begin offering training workshops to teach managers how to identify potential change agents on their teams, and how to nurture their growth and professional development.

5.2. FORMALIZE RELATIONSHIPS THAT CHANGE AGENTS HAVE INFORMALLY BUILT

Change agents excel at building bridges of communication **among people who normally wouldn't interact with one another.** These bridges are vital for fostering mutual respect and trust, gaining wider input into the design of workflows and other solutions, and eliminating pain points that are hindering a team's ability to adapt to change. You want to formalize these relationships within your organization, to ensure that other teams and departments can benefit from these enhanced levels of interaction. This process starts by identifying and documenting the relationships that change agents have built. Then, you want to use your management-level authority to memorialize these relationships as official workflows and protocols that get adopted by the entire organization.

5.3. HELP CHANGE AGENTS GET IN FRONT OF THE C-SUITE

One of the key reasons that change agents are effective in helping to drive change among their peers is because they have strong communication skills. They understand how to translate technical specs into productivity, and how to translate big-picture corporate change mandates into tangible, employee-centric messages that resonate with others. It's important to help change agents recognize their effectiveness at communicating by getting them and their ideas in front of the C-suite. Your

On their own, change agents are not likely to think of themselves as valuable to the C-suite; it's your job to help them appreciate all they have to offer.

C-suite may want to use change agents as a sounding board prior to rolling out change initiatives, and/or may directly want to solicit suggestions and ideas from change agents. On their own, change agents are not likely to think of themselves as valuable to the C-suite; it's your job to help them appreciate all they have to offer. You can help them maximize their effectiveness in front of the C-suite by giving them a place to prepare. You can create internal chat rooms and sandboxes that bring your change agents together to flesh out their ideas, rate them, and decide how and when to put these ideas in front of the C-suite.

5.4. HELP CHANGE AGENTS SPARK VIRAL CHANGE

Because change agents don't have formal mandates and workflows for initiating change, their successes with change come because they know how to make change "go viral." When change agents spark change, they intuitively realize they don't have management-level drivers standing behind them to compel this change. Instead, change agents seek to build grassroots momentum for change, where employees start embracing change because of the change agent's nurturing, encouraging approach. You want to support these approaches by empowering change agents to spark viral change. You can introduce them to employee communications tools already in use, as well as help them identify and acquire additional "viral" tools.

You also can approve them to participate in outside tutorials and other professional development that help them make optimal use of these tools.

5.5. GUIDE CHANGE AGENTS IN NAVIGATING THE BUREAUCRACY OF CHANGE

The C-suite is trained to be highly deliberative and deeply cautious about greenlighting projects. Senior-level executives will typically ask many questions, express concerns openly, and want more research before announcing their decisions. This approach tends to be incompatible with how change agents think and act; they develop ad-hoc solutions to problems through trial and error and gut instinct, and don't understand or want to deal with bureaucracy.

To avoid letting your change agents become frustrated and disenchanted by endless meetings and protracted approvals processes, you can help them navigate the bureaucracy of change, and advise them how to efficiently navigate internal company politics. For example, if your change agents come to you with an ambitious idea, you can help them create a small pilot project first, develop C-suite-friendly KPIs to track its effectiveness, and help them prepare a presentation to the C-suite on the project's outcomes. Afterward, you can hold the C-suite accountable to whatever actions they've committed to, and guide your change agents in how to help implement these actions.

Key Talking Points to Cover with Your Colleagues

As you look to optimally position your change agents to drive meaningful change within your organization, remember that change agents often don't even recognize they are change agents. It's your job to nurture their talents, help them realize their potential, and connect them with other like-minded individuals to create strong internal networks of change agents.

- Change agents need institutional processes and systems in place to optimally work their transformative magic.

- It's your job to help change agents navigate the bureaucracy of change, and ensure they can get their ideas effectively in front of the C-suite.

- Remember to assist change agents in extending the reach of their changes and getting their changes to go viral.

PILLAR 6

Overcoming Obstacles that Threaten Momentum

MEET GARRETT

Garrett is the CEO of a global engineering company that began implementing ServiceNow three years ago to transform enterprise services and boost profitability. The IT Service Management (ITSM) component largely went off without a hitch, and employees have become dependent on features like the self-service portal for managing their IT-related requests. Each subsequent ServiceNow component, including Project Portfolio Management (PPM), seemed to start off strong, but ultimately has floundered. In particular, Garrett stopped seeing the unified, consistently strong forward momentum that the company was experiencing during ITSM implementation. Garrett sees that other strategic business initiatives and goals have taken priority, and employees have been reverting back to some of their old habits and workflows. Indeed, Garrett has repeatedly heard grumbling about all of the time his company has spent trying to fix workflows and systems that many employees say were never broken. Garrett still believes in ServiceNow and wants to see ServiceNow continue to work its transformative magic across the organization, just as ServiceNow did with the company's ITSM. Garrett talks over these issues with his C-suite, and they all agree they'll need to up their game to sustain ServiceNow's forward momentum. Garrett's next step is figuring out what strategies and approaches he can implement to turn around the ship and make ServiceNow the success it should be.

The less-than-stellar results that Garrett experienced with digital transformation are commonplace among organizations across all industries. A 2018 study by the market intelligence firm International Data Corporation found that 69% of enterprise companies believe their digital transformation initiatives have been less than successful in meeting stated goals, and only 19% believe they're truly innovating as a result of their digital transformation initiatives. The reality is that digital transformation is more of a journey and practice than a destination. At its core, digital transformation depends entirely on individuals who are not always amenable to change, and who may be inclined to give up on change over time.

You may want to believe that you can achieve ServiceNow digital transformation just by getting a majority on board and that you can simply power through any resistance you encounter from employees. Unfortunately, this will never be the case, and that's because successful digital transformation initiatives take a village. If every employee isn't moving in unison toward change, you will create silos in your organization that, over time, will derail your momentum and stunt your progress. Every employee needs to feel empowered, valued, and confident that their ideas and strategies are being heard, incorporated, and efficiently utilized. Let's explore best-practices strategies for overcoming the obstacles that interrupt and derail organizational change.

A 2018 INTERNATIONAL DATA CORPORATION STUDY

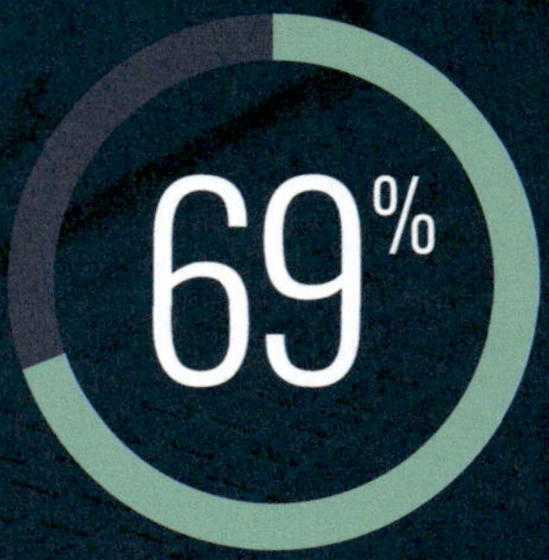

of enterprise companies believe their digital transformation initiatives have been less than successful in meeting stated goals.

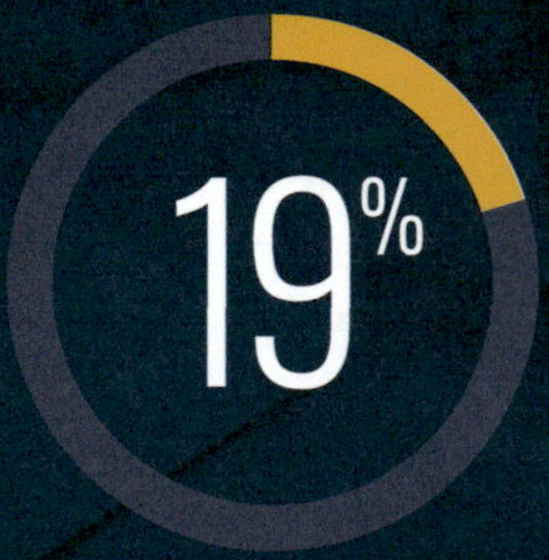

of enterprise companies believe they're innovating as a result of their digital transformation initiatives.

6.1. RECOGNIZE THAT EMPLOYEES RESIST CHANGE FOR MANY REASONS

Employees resist change for many reasons; thus, the first step in managing employee resistance is to acknowledge that this resistance is real. Some employees are inherently averse to risk and have a compelling need to protect the systems and jobs that are working. Other employees lack self-confidence or have a generalized fear of the unknown. Still others are going through unrelated personal issues in their own lives that confound their ability to deal with workplace change. When you take the time to learn why employees resist change, you can identify effective interventions to help them through this process:

- If employees lack confidence, you can connect them to more training and resources.
- If employees are averse to risk, you can slow down the pace of change.
- If employees are dealing with personal circumstances, you can work with them to identify mutually agreeable alternative pathways and timelines to achieve change.

6.2. COURT YOUR NAYSAYERS AND CRITICS

The worst thing that you can do is to ignore critics or to assume that naysayers will eventually come around and embrace change. Over time, they are likely to become hardened in their beliefs toward the change and pose an even more formidable obstacle to change down the road. The most effective solution is to encourage them to air their concerns at the upfront; indeed, you want to be talking them through each of the areas and issues that are of most concern to them. At the end, you want to invite them to participate in the change process, even if in a limited capacity on their own terms. Participating in the change is likely to change their attitudes and views toward the change and, at the very least, lessen their resolve to actively fight the change.

6.3. BE A GOOD LISTENER, RESPOND LATER

When you don't give employees a proper forum to express their concerns, frustrations, and grievances about change, they become more likely to resist change. You want to work proactively and earnestly to listen and absorb employee feedback; you especially don't want to talk over them or cut off their responses. In fact, the best way to proceed is to always plan on having two conversations. In the first conversation, you simply listen, responding only to direct questions asked of you and ending the conversation by saying you will research and follow up on what you've heard. In the second conversation, you want to summarize what they told you—a powerful cue to show them you listened during the first conversation—and then explain what you will or won't be doing differently as a result. If you still encounter hesitancy and resistance after the second conversation, repeat the same pair of conversations again.

> The worst thing that you can do is to ignore critics or to assume that naysayers will eventually come around and embrace change. Over time, they are likely to become hardened in their beliefs toward the change, and to pose an even more formidable obstacle to change down the road.

6.4. GET EXECUTIVE SPONSORSHIP AND CREATE A STEERING COMMITTEE

Despite everyone's best intentions, momentum for change is difficult to sustain over the long term. Other priorities come up that cause a company's focus to drift. Employees start to question and doubt how much C-suite support is behind a change initiative. And as other priorities and initiatives are rolled out over time, employees' mental bandwidth is stretched thin. The solution is to get executive sponsorship for your ServiceNow digital transformation initiative, as well as create a steering committee that is specifically charged with keeping it on track. Delegating explicit authority to manage change will ensure your organization has a long-term advocate for digital transformation—a ServiceNow champion that will impress upon everyone that the change initiative is still an organizational priority.

6.5. ENSURE MANAGEMENT IS MODELING THE CHANGE YOU SEEK

Every day, employees juggle many competing demands on their time. They are continually looking to their managers for guidance on what tasks and projects are most important to prioritize as a way to fast-track progress. Management should, of course, continually tell employees what's important. But beyond that, one of the most effective ways to ensure digital transformation stays on an employee's priority list is for managers to exemplify what transformation looks like in their own work. Managers should visibly demonstrate through their actions that they are 100% committed to digital transformation and that digital transformation is a top priority that will improve the employee experience. Employees are continuously watching managers for cues about what's really important to the organization and will emulate behaviors that they view as appreciated and valued by the organization. You ultimately want to achieve 100% alignment between what managers say and what they do.

Key Talking Points to Cover with Your Colleagues

Obstacles in your path to ServiceNow digital transformation are not roadblocks you can simply ignore. To sustain change over the long term, you must be constantly on the offensive, identifying obstacles that you proactively work to overcome.

- Employees resist change for many reasons; it's your job to understand why and to work with them to identify solutions.

- Your naysayers and critics can become among your most powerful allies if you truly listen to them.

- Management plays a key role in helping you overcome obstacles; when managers resist change or don't exemplify the change, they become just as much of an obstacle to change as any other resistant employee.

73% VS 46%

In companies undergoing successful digital transformation, 73% report visibly engaged firstline workers, compared to 46% among companies struggling with digital transformation.

2017 MCKINSEY & COMPANY SURVEY

43% VS 21%

A Forbes Insights study found that 43% of companies with a large portion of digitally empowered firstline workers report excellent progress on digital transformation. By contrast, only 21% of companies with few digitally empowered firstline workers report comparable progress. Similarly, 34% of companies with highly empowered firstline workers report excellent progress leveraging data analytics, compared to 14% among low-empowerment organizations.

2018 FORBES INSIGHTS

For more findings, read "Going Digital? Your Firstline Workers Will Pave Your Way To Success" from Forbes.com.

PILLAR 7

Measuring the Effectiveness of Change

MEET TAYLOR

Taylor is the executive vice president for global operations at a commercial investment bank. Taylor has been delegated responsibility for managing the implementation of ServiceNow across the organization, beginning with IT Service Management (ITSM) and continuing with Customer Service Management (CSM) and Project Portfolio Management (PPM). The digital transformation initiative is massive, with Taylor at the center working to coordinate ServiceNow's technical buildout, development of an implementation roadmap, and change management. Taylor has been trained her entire career to think about enterprise initiatives in terms of KPIs (Key Performance Indicators), and yet she feels too overwhelmed at the moment to focus on developing KPIs for ServiceNow. She knows that as soon as she gets a few more major items off her to-do list, she'll be able to start pulling numbers and tracking how employees are responding to ServiceNow. But before she can come up for air, her CEO stops by her office at the end of a hectic day to tell her that he just came across a report showing that ServiceNow can deliver a 195% return on investment (ROI) in just nine months. Taylor panics and realizes she can no longer put off developing KPIs to measure the effectiveness of ServiceNow digital transformation.

The wake-up call that Taylor received is unfortunately commonplace among organizations struggling to manage the many moving parts of ServiceNow implementation. A 2017 Gartner survey of business CEOs indicated that only about half of organizations have KPIs in place to measure success with their digital transformation initiatives. Even if the technology behind a transformation initiative is running smoothly, there's always a big unknown regarding how quickly and efficiently employees will respond to the change. To have confidence you have been successful with your organizational change management efforts, focus on creating comprehensive, standardized best-practices methods and metrics for measuring the effectiveness of change—beyond just simple calculations of

ROI. Let's explore some of the most important considerations to keep in mind as you develop your approach to assessing progress toward your digital transformation goals.

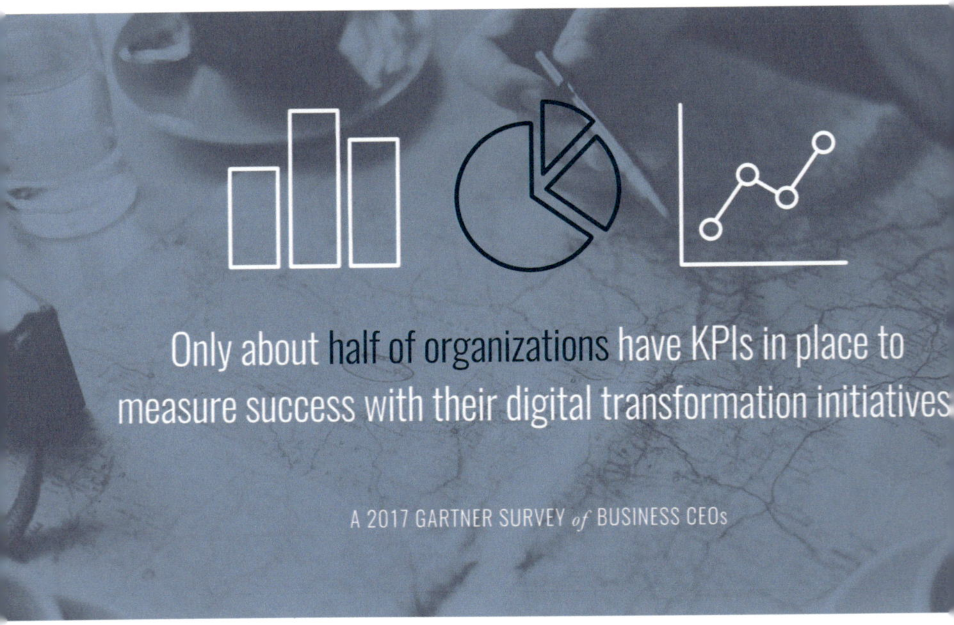

7.1. DEVELOP YOUR METRICS AT THE UPFRONT

By the time you start down the path of digital transformation, you should already have your KPIs in place. You want to be able to track progress through implementation, so you can confidently report on progress and celebrate wins. Examples of common ServiceNow KPIs include

» resolution rate on high-priority incidents,
» resolution rate for incidents on first assignment,
» rate of reopened incidents,
» average time to resolve a high-priority incident, and
» average number of incidents created per user.

7.2. DEVELOP QUALITATIVE KPIS THAT CAN BE TRACKED SOONER THAN ROI

The classic way to calculate ROI is to divide the net financial benefit of the investment by the amount invested. This calculation is simple, but not readily available to you at the early stages of change management. You cannot afford to wait until you can calculate ROI to know that your change initiative is working. That's why you need to develop KPIs that will give you much earlier indications of success. One particularly effective type of KPI is measuring your employees' progress in adapting to change. Employee KPIs can be developed around

- ability to articulate the goals and rationale for change,
- proficiency with a new system or workflow,
- aptitude for change, and
- level of buy-in and support for change.

Significantly, it's important to remember that these qualitative measures are constantly evolving and cannot be captured retroactively; you must have already developed KPIs at the upfront to capture these snapshots in time.

7.3. CREATE MULTIPLE KPIS TO MEASURE THE SAME METRIC

Because many of the first KPIs you'll gather will measure progress qualitatively, you should consider creating multiple overlapping KPIs that measure the same metric. This strategy will increase C-suite confidence in the results. For example, to track level of employee engagement in a change initiative, you don't just want to document how employees are performing through observation. You also want to solicit performance feedback from their supervisors. Similarly, if they play any role in shaping the customer experience, you want to compile data on customer and partner satisfaction.

7.4. DEVELOP A PUBLICLY AVAILABLE KPI SCORECARD

Digital transformation affects the entire organization, and for it to be successful, everyone must buy into and support digital transformation. Thus, you want to create full transparency around how the organization is performing; this is most effectively accomplished by creating a KPI scorecard that is publicly available and updated in real time. This scorecard will encourage employees to both celebrate wins and to quickly recognize when they're underperforming and must make a change.

7.5. REACH CONSENSUS IN ADVANCE ON WHEN YOU WILL REVERSE COURSE

KPIs are crucial for monitoring progress on a digital transformation journey, but what happens when the KPIs indicate less-than-stellar results? You'll need to know what are the thresholds at which the organization will reverse course and/or implement an intervention. Because KPIs only illuminate a problem and don't offer clear pathways for fixing problems, you don't want to get into a protracted debate about if and how you should reverse course.

That's why it's important to reach consensus in advance on what the thresholds of tolerance are: when you hit a threshold, there will be no debate or disagreement on how you'll proceed. For example, if KPIs indicate that employees are not adapting to new workflows and systems, you must agree on how underperforming your employees must be for you to intervene. And if you decide to intervene, you must agree on the threshold at which you will implement, for example, additional training vs. the threshold at which you will begin reflecting performance shortcomings in an employee's performance review.

Key Talking Points to Cover with Your Colleagues

KPIs are crucial to tracking progress toward your change initiatives. To ensure KPIs provide insights that you can rely on, it's your responsibility to design KPIs that are consistent and comprehensive.

- Qualitative KPIs of employee performance are especially important to develop at the upfront.

- You can create overlapping KPIs that measure the same metric.

- A publicly accessible KPI scorecard is the best way to keep employees on track.

- To avoid a dispute later about how to interpret KPIs, reach consensus in advance on how you'll interpret outcomes.

FINAL THOUGHTS

Fostering a Perpetual Change Culture

There's no question that creating a rock-solid strategy for digital transformation is critical to achieving a strong return on your ServiceNow investment. But sustaining your progress over the long term is equally important—even as it remains elusive for many organizations. When organizations fail to achieve the ServiceNow transformation they seek, they can often trace the shortcomings back to inadequate processes and standards that were not replicated and/or were not sufficiently rigorous. An organization may think it's put in place bulletproof OCM protocols and processes to support ServiceNow implementation, but often these safeguards simply aren't rigorous enough to stand the test of time.

The only way to guarantee long-term success with ServiceNow digital transformation is to never let up. Every leader, from the C-suite to front-line management, must remain 100% committed to change, to live and breathe digital transformation.

Moreover, OCM leaders must recognize that the path of least resistance is always to fall back on old habits and to lose focus over time in the face of competing priorities. Thus, it's critical that OCM leaders never think that ServiceNow-powered digital transformation can "manage itself." The OCM managers who are most likely to accomplish their long-term ServiceNow goals are the ones who are continually questioning their progress, looking for fresh data to benchmark performance, and rolling out new approaches to help keep OCM front and center across the organization. The best OCM leaders also are always in listening and observing mode, seeking out suggestions and ideas from employees on how to more effectively manage change.

Although OCM is a long and, at times, unpredictable journey, you don't need to go this journey alone. You can turn to seasoned consultants who specialize in managing the massive organizational change that accompanies ServiceNow implementation. Crossfuze, a global leader in ServiceNow implementations, helps companies every day with much more than just the technical rollout of ServiceNow; our experienced consultants are masters at helping organizations develop comprehensive, customized plans for managing enterprise-wide change.

If you'd like help with OCM to achieve rapid, strong returns on your ServiceNow investment, please reach out to us at LetsTalk@crossfuze.com. We'd be delighted to help you create your ServiceNow success story.

Crossfuze
Enterprise Services That Win

Made in the USA
Monee, IL
05 November 2026

67384981R00040